HARMONOGRAPH

*Unison (1:1): a spiral, a spiral drawn the same way over a
spiral, and a spiral drawn the opposite way over a spiral.*

Originally published in Wales by Wooden Books Ltd. in 2003;
first published in the United States of America in 2003 by
Walker Publishing Company, Inc.

Published simultaneously in Canada by Fitzhenry and Whiteside,
Markham, Ontario L3R 4T8

Printed on recycled paper

For information about permission to reproduce selections from this
book, write to Permissions, Walker & Company, 435 Hudson Street,
New York, New York 10014

Library of Congress Cataloging-in-Publication Data

Ashton, Anthony.
Harmonograph : a visual guide to the mathematics of music /
Anthony Ashton.
p.   cm.
ISBN 0-8027-1409-9 (alk. paper)
1. Harmonographs.   2. Music theory—Mathematics.   I. Title.
QC228.3.A78   2003
781.2—dc21   2003042262

Visit Walker & Company's Web site at www.walkerbooks.com

Printed in the United States of America

2 4 6 8 10 9 7 5 3 1

HARMONOGRAPH

A VISUAL GUIDE TO THE MATHEMATICS OF MUSIC

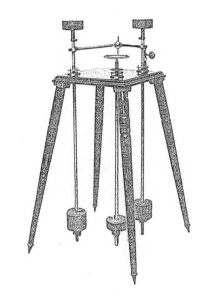

Anthony Ashton

Walker & Company
New York

Dedicated to John, Antonia, and Imogen

Grateful acknowledgment to Harmonic Vibrations and Vibration Figures, *by Joseph Goold, Charles E. Benham, Richard Kerr, and L. R. Wilberforce, edited by Herbert C. Newton, Newton & Co., 1909;* Science and Music, *by Sir James Jeans, Cambridge, 1937;* Sound, *by John Tyndall, Appleton & Co., 1871;* Les Recreations Scientifiques, *by Gaston Tissandier, Masson, 1881. The image on page 47 is taken from the extraordinary book* Cymatics: A Study of Wave Phenomena and Vibration, *by Hans Jenny, © 2001, MACROmedia Publishing, and used here by kind permission. I am particularly grateful to my grandson for help with the intricacies of musical theory.*

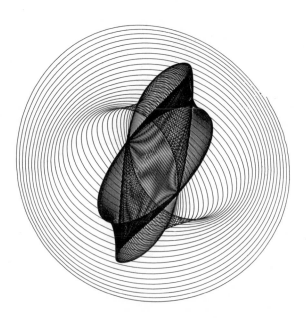

CONTENTS

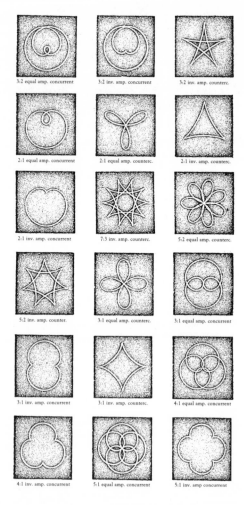

Harmonic patterns from Sir Thomas Bazley's Index to the Geometric Chuck *(1875), showing concurrent and countercurrent phases with equal and inverted amplitudes.*

INTRODUCTION

Many of the drawings in this book were produced by a simple scientific instrument known as a harmonograph, an invention attributed to a Professor Blackburn in 1844. Toward the end of the nineteenth century these instruments seem to have been in vogue. Victorian gentlemen and ladies would attend soirées or *conversazioni*, gathering around the instruments and exclaiming in wonder as they watched the beautiful and mysterious drawings appear. A shop in London sold portable models that could be folded into a case and taken to a party. There may well be some of these instruments hidden in attics all over the world.

From the moment I first saw drawings of this kind I was hooked. Not only because of their strange beauty, but because they seemed to have a meaning—a meaning that became clearer and deeper as I found out how to make and operate a harmonograph. The instrument draws pictures of musical harmonies, linking sight and sound.

However, before going any further I feel I should issue a health warning. If you too are tempted to follow this path, beware! It is both fascinating and time-consuming.

I have acknowledged my debt to the book *Harmonic Vibrations*. It was coming across this book in a library soon after the end of the second world war that introduced me to the harmonograph. Seeing that the book had been published by a firm of scientific instrument makers on Wigmore Street I went one day to see if they were still there. They were, though reduced merely to making and selling projectors. I went into the shop and held up my library copy of the book for the elderly man behind the counter to see.

"Have you any copies of this book left?" I asked him.

He stared at me as though I were some sort of ghost and shuffled away without a word, returning in a few minutes with a dusty, unbound copy of the book.

"That's marvelous," I said, "how much do you want for it?"

"Take it," he said, "it's our last copy, and we're closing down tomorrow."

So I have always felt that someday I must write this book.

Girton, 2002

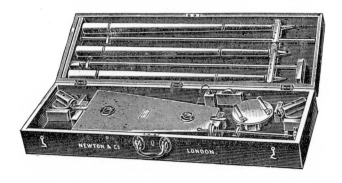

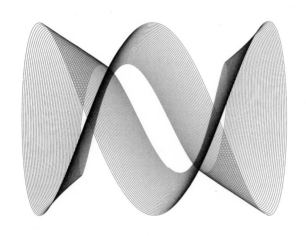

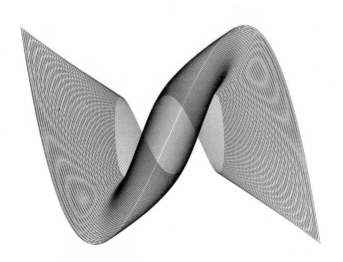

The Discovery of Harmony
on passing a blacksmith

To understand what the harmonograph does we need first to glance at the elements of musical theory.

Pythagoras, some 2,500 years ago, is credited with discovering that the pleasing experience of musical harmony comes when the ratio of the frequencies consists of simple numbers. A tale tells how while taking a walk he passed a blacksmith's shop. Hearing familiar harmonies in the ringing tones of the hammers on the anvil, he went in and was able to determine it was the weights of the hammers that were responsible for the relative notes.

A hammer weighing half as much as another sounded a note twice as high: an *octave* (2:1). A pair weighing 3:2 sounded beautiful, a *fifth* apart. Simple ratios made appealing sounds.

The pictures opposite show experiments the philosopher went on to make (from Gafurio's *Theorica Musice*, 1492), as he found that all simple musical instruments work in much the same way, whether they are struck, plucked, or blown.

Deeply impressed by this link between music and number, Pythagoras drew the metaphysical conclusion that all nature consists of harmony arising from number, precursor to the modern physicist's assumption that nature conforms to laws expressed in mathematical form. Looking at the pictures you will see that in every example—hammers, bells, cups, weights, or pipes—the same numbers appear: 16, 12, 9, 8, 6, and 4. These numbers can be paired in quite a few ways, all of them pleasant to the ear, and, as we shall see, also pleasant to the eye.

THE MONOCHORD OF CREATION
a singular string theory

There are seven octaves in the keyboard of a piano and nearly eleven in the total range of sound heard by the average person. The highest note of each octave has a frequency twice that of the first so the frequencies increase exponentially, on a scale beginning at 16 oscillations per second (16 *Hertz*) with the lowest organ note and ending with about 20,000 per second. Below 16 Hertz we experience rhythm. A range of ten octaves represents about a thousandfold increase in frequency ($2^{10} \approx 10^3$).

There is a hint here of what we can think of as the great monochord of the universe, also on a scale, this time stretching from a single quantum fluctuation at the bottom, to the observable universe at the top, passing through the various "octaves" of atom; molecule; quantities of solid, liquid, and gaseous matter; creatures great and small; planets; stars; and galaxies. Here too the scale is exponential, but usually measured in powers of ten, and covering a range of more than 10^{40}.

Robert Fludd's seventeenth-century engraving (*opposite*) tells a similar story: The musical scale follows the same exponential principle underlying the design of the universe.

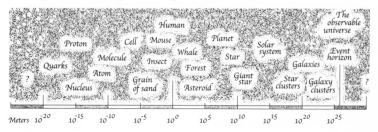

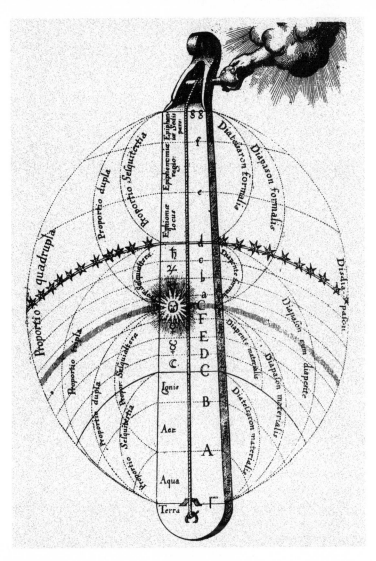

OVERTONES AND INTERVALS
harmonic ratios in and outside the octave

How are musical scales constructed? Listen very carefully as you pluck a string and you will hear not only the main note, or *tonic*, but also a multitude of other harmonics, the *overtones*.

The principle is one of harmonic resonance, and affects not only strings and ringing hammers, but columns of air and plates too. Touching a string with a feather at the halfway or third point, as shown below, encourages regularly spaced stationary points, called *nodes*, and an overtone can be produced by bowing the shorter side. The first three overtones are shown opposite (*top*).

Musicians, however, need notes with intervals a little closer together than the overtone series, which harmonize within an octave. The lower diagram opposite shows the overtone series on the left, and the intervals developing within the octave on the right, in order of increasing dissonance, or complexity.

"All discord harmony not understood" wrote Alexander Pope. The brain seems to grasp easily the relationships implicit in simple harmonies, an achievement bringing pleasure; but with increasing complexity it falters and then fails, and failure is always unpleasant. For most people enjoyment fades as discord increases, toward the end of the series opposite. And, as we shall see, that is where the harmonograph drawings fade too.

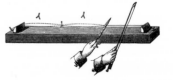

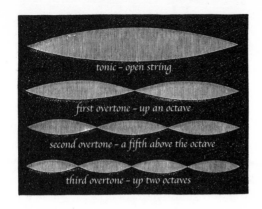

tonic – open string

first overtone – up an octave

second overtone – a fifth above the octave

third overtone – up two octaves

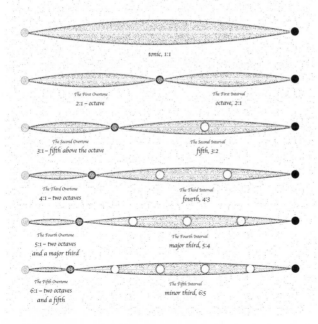

tonic, 1:1

The First Overtone
2:1 – octave

The First Interval
octave, 2:1

The Second Overtone
3:1 – fifth above the octave

The Second Interval
fifth, 3:2

The Third Overtone
4:1 – two octaves

The Third Interval
fourth, 4:3

The Fourth Overtone
5:1 – two octaves
and a major third

The Fourth Interval
major third, 5:4

The Fifth Overtone
6:1 – two octaves
and a fifth

The Fifth Interval
minor third, 6:5

WHOLE TONES AND HALFTONES
the fifth and the octave get their names

Pythagoras's hammers hide a set of relationships dominated by octaves (2:1), fifths (3:2), and fourths (4:3). The fifth and fourth combine to make an octave ($3:2 \times 4:3 = 2:1$), and the difference between them ($3:2 \div 4:3$) is called a *whole tone*, value 9:8.

A natural pattern quickly evolves, producing seven discrete nodes (or notes), separated by two *halftones* and five whole tones, like the sun, moon, and five planets of the ancient world.

The fifth (3:2) naturally divides into a major third and minor third ($3:2 = 5:4 \times 6:5$), the major third basically consisting of two whole tones, and the minor third of a whole tone and a halftone. The thirds can be placed major before minor (*to give the major scale shown in the third row, opposite*) or in other ways.

Depending on your harmonic moves, or *melody*, different *tunings* appear, for example two perfect whole tones ($9:8 \times 9:8 = 81:64$) are not in fact the perfect major third 5:4, but are slightly sharp as 81:80 (the *syntonic* or *synoptic comma*, the Indian *sruti*, or *comma of Didymus*), which will be discussed more later.

Simple ratios, the octave and fifth, have given rise to a basic scale, a pattern of whole tones and halftones and, depending on where in the sequence you call home, seven *modes* are possible.

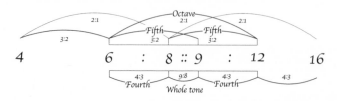

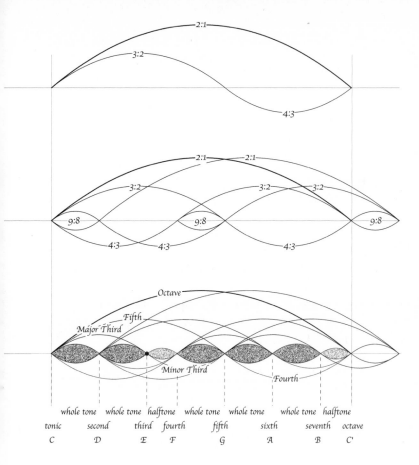

The basic manifestation of the scale. In Pythagorean tuning all whole tones are exactly 9:8, creating the leimma halftone of 256:243 between its major third (81:64) and the perfect fourth (4:3). The sixth and the seventh are defined as successive perfect whole tones above the fifth.

In Diatonic tuning the major third is perfect at 5:4, which squeezes the second whole tone to 10:9 (a minor whole tone), leaving 16:15 as the diatonic halftone up to the fourth. The diatonic sixth is 5:3, a major third above the fourth, a minor whole tone above the fifth. The diatonic seventh (15:8) is a major whole tone above that, a major third above the fifth and a halftone below the octave.

11

ARRANGING THE HARMONIES
the power of silence

The simple ratios of the primary overtones and undertones can be plotted on an ancient grid known as a *lambdoma* (*opposite, top*), after the greek letter λ. Some intervals are the same (8:4 = 6:3 = 4:2 = 2:1), and if lines are drawn through these it quickly becomes apparent that the identities converge on the silent and mysterious ratio 0:0, which is outside the diagram.

A further contemplative device used by the Pythagoreans was the *Tetraktys*, a triangle of ten elements arranged in four rows (1+2+3+4=10). The basic form is given opposite, lower left, the first three rows producing the simple intervals. In another lambdoma (*opposite, lower right*), numbers are doubled down the left side and tripled down the right, creating tones horizontally separated from their neighbors by perfect fifths. After the trinity (1, 2, and 3) notice the numbers produced, 4, 6, 8, 9, 12, and then look again at the picture on page 5.

Below are interval positions on a monochord.

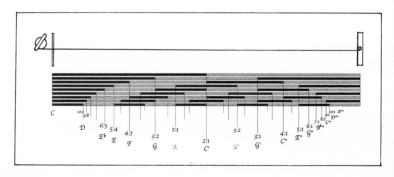

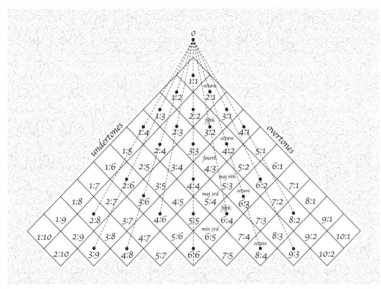

Pythagorean and medieval tunings, called 3-limit, recognized no true intervals except for ratios involving 1, 2, and 3. The lambdoma below, right expresses this numerically as any element relates to any neighbor by ratios only involving 1, 2, and 3, so we can move around by octaves and fifths. Squares ($4=2^2$, $9=3^2$) and cubic volumes ($8=2^3$, $27=3^3$) also appear. Add further rows and the numbers for the Pythagorean scale soon appear—1 9:8 64:81 4:3 3:2 27:16 16:9 2:1. This has four fifths and five fourths but no perfect thirds or sixths. These came later with the diatonic scale and its perfect thirds (6:5:4) as polyphony and chords slowly took over from plainchant and drone.

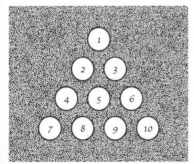

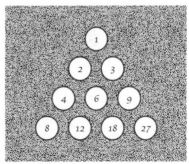

LISSAJOUS FIGURES
sound made shape

In the mid-nineteenth century, Jules Lissajous, a French mathematician, devised an experiment: He found that if a small mirror was placed at the tip of a tuning fork, and a light beam was aimed at it, the vibration could be thrown onto a dark screen. When the tuning fork was struck, a small vertical line was produced, and if quickly cast sideways with another mirror it produced a sine wave (*below*).

Lissajous wondered what would happen if instead of casting the wave sideways he were to place another tuning fork at right angles to the first to give the lateral motion. He found that tuning forks with relative frequencies in simple ratios produced beautiful shapes, now known as Lissajous figures.

On the screen (*opposite, top*), we see the octave (2:1) as a figure eight, and below it various *phases* of the major and minor third. These were some of the first fleeting pictures of harmony, which were doubtless familiar to Professor Blackburn when he devised the harmonograph.

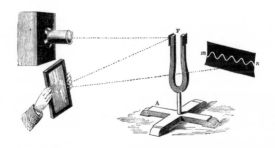

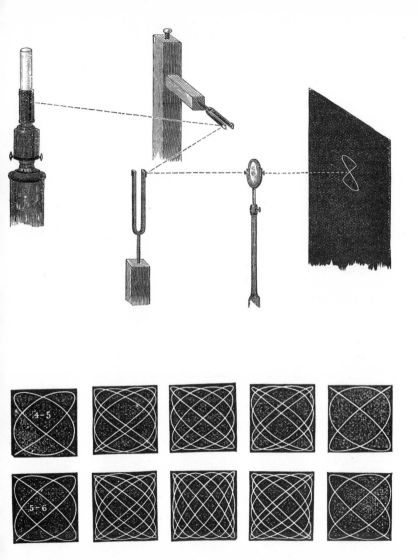

15

THE PENDULUM
keeping time

A fundamental law of physics (in one formulation) states that left to itself any closed system will always change toward a state of equilibrium from which no further change is possible.

A pendulum is a good example. Pulled off center to start, it is in a state of extreme disequilibrium. Released, the momentum of its swing carries it through to nearly the same point on the other side. As it swings it loses energy in the form of heat from friction at the fulcrum and brushing against the air. Eventually the pendulum runs down, finally coming to rest in a state of equilibrium at the center of its swing.

Going back 500 years, Galileo, watching a swinging lamp in the cathedral of Pisa, realized the frequency of a pendulum's beat depends on its length: The longer the pendulum the lower the frequency. So the frequency can be varied at will by fixing the weight at different heights. Most important, as the pendulum runs down, the frequency stays the same.

Here, therefore, is a perfect way to represent a musical tone, slowed down by a factor of about a thousand to the level of human visual perception. For a simple harmonograph two pendulums are used to represent a harmony, one with the weight kept at its lowest point, while the weight on the other is moved to wherever it will produce the required ratio.

As we shall see, the harmonograph combines these two vibrations into a single drawing, just as two musical tones sounded together produce a single complex sound.

The theoretical length of the variable pendulum that will produce each harmony can be calculated, for the frequency of a pendulum varies inversely with the square root of its length. This means that while the frequency doubles within the octave, the length of the pendulum is reduced by a factor of four.

Figures are given for a pendulum 32 inches (80 cm.) long, a convenient length for a harmonograph. These theoretical markers provide useful "sighting shots" for most of the harmonies. Note that the pendulum length is measured from the fulcrum to the center of the weight.

Interval Name	Approx. Note	Diatonic Ratio	Length (cm)	Freq. (s¹)
Octave	C'	2:1	20	66.0
Maj. 7th	B	15:8	22.8	62.8
Min. 7th	B♭	9:5	24.7	59.4
Maj. 6th	A	5:3	28.8	55.8
Min. 6th	G♯	8:5	31.2	53.6
5th	G	3:2	35.6	50.3
4th	F	4:3	45.0	44.7
Maj. 3rd	E	5:4	51.2	41.9
Min. 3rd	E♭	6:5	55.6	40.2
2nd	D	9:8	63.2	37.7
Halftone	C♯	16:15	70.3	35.7
Unison	C	1:1	80	33

When a pendulum is pulled back and then released, the weight tries to fall toward the center of the Earth, accelerating as it does so. As the pendulum runs down, the rate of acceleration, and so the speed of travel, is reduced, but in equal proportion to the distance of travel.

The result is that the period (the time taken for two beats) or the number of periods in a given unit of time (the frequency) remains unchanged. In the picture to the left the frequencies of beats x and y are the same.

For the pendulum formula, see page 53.

17

Two Harmonographs
lateral and rotary

In the simplest version of the harmonograph two pendulums are suspended through holes in a table, swinging at right angles to one another. Projected above the table, the shaft of one pendulum carries a platform with a piece of paper clipped to it, while the shaft of the other pendulum carries an arm with a pen.

As the pendulums swing, the pen makes a drawing that is the result of their combined motion (*opposite, left side*). Both pendulums begin with the same length, one is then shortened by sliding the weight upward and securing it with a clamp at various points. The harmonic ratios can be displayed in turn.

By using three pendulums, however, two circular, or rotary, movements can be combined, with fascinating results (*opposite, right side*). Two of the pendulums swing at right angles as before, but are now both connected by arms to the pen, which in all rotary designs describes a simple circle.

Situated under the circling pen, the third and variable pendulum is mounted on gimbals, a device familiar to anyone who has had to use a compass or cooking stove at sea. Here it acts as a rotary bearing, enabling the pendulum carrying the table to swing in a second circle under the pen. As the pen is lowered the two circles are combined on the paper.

A further source of variation is also introduced here, for the two circular motions can swing in the same (concurrent) or opposite (countercurrent) directions, producing drawings with very different characteristics.

18

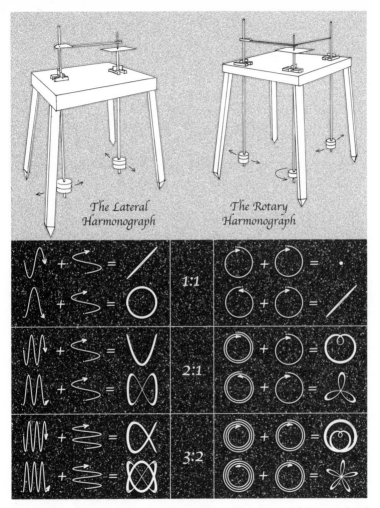

Two harmonographs and some of the simple patterns they draw. On the left the simple lateral version and its patterns (open and closed phase); on the right the three-pendulum, rotary harmonograph and its drawings (concurrent and countercurrent).

SIMPLE UNISON—1:1
and the arrow of time

The simplest harmonograph drawing is produced when both pendulums are the same length and the table is stationary. With the pen held off the paper both pendulums are pulled back to their highest points. One is released, followed by the other when the first is at its midpoint. The pen is then lowered onto the paper to produce a circle developing into a single spiral.

If the two pendulums are released together then the result will be a straight diagonal line across the paper, the "closed" phase of the harmony, as opposed to the circular "open" phase. At intermediate phase points elliptical forms appear (*below*).

The running-down of harmonograph pendulums is an exact parallel to the fading of musical notes produced by plucked strings, and can also be thought of as graphically representing the "arrow of time" (*opposite*), with the unchanging ratios of the frequencies representing the eternal character of natural law. The characteristics of the drawings result from the meeting of the running-down process with the laws represented by the various frequency ratios. We see that music, like the world, is formed from unchanging mathematical principles deployed in time, creating complexity, variety, and beauty.

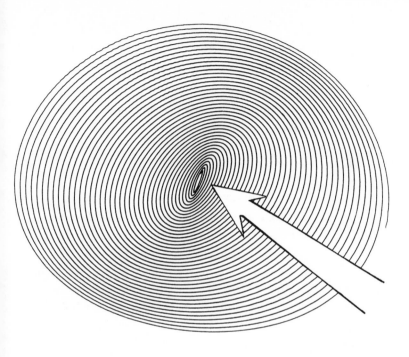

The inexorable direction of change, linked to the asymmetry of time (before-now-after), was vividly described by the scientist Arthur Eddington (1882–1944) as "the arrow of time." Throughout the process of continuing universal degradation, the dwindling stock of useful energy encounters a hierarchy of fixed physical laws conforming to mathematical formulas, and from the interaction of these unchanging laws with the arrow of time comes a changing world of astonishing complexity, variety, and beauty. The pendulum runs down from a state of disequilibrium to one of equilibrium, and the same is true, we are told, of the universe, the ultimate closed system. From a state of extreme disequilibrium it plunged via the Big Bang toward its future ultimate state of utterly dark, frozen equilibrium. Between the beginning and the end there is a continual, cumulative transformation of useful energy, capable of forming temporary structures and causing events, into useless energy forever lost.

NEAR UNISON
lateral phases and beat frequencies

A source of pleasing variety in harmonograph drawings comes from small departures from perfect harmonies. This seems to involve a principle widespread in nature as well as in the work of many artists. There is a particular charm in the near miss.

An example from music suggests itself here. When two notes are sounded in near unison, the slight difference in their frequencies can often add richness or character to the sound. The two reeds producing a single note in a piano accordian have slightly different frequencies, the small departure from unison causing *beats*, a warbling or throbbing sound (*see page 53*).

Set the weights for unison and then shorten the variable pendulum slightly. Swing the pendulums in open phase, producing a circle turning into an increasingly narrow ellipse and then a line. If the pen is allowed to continue, the line will change into a widening ellipse, a circle, and a line again at right angles to the first. And so on. The instrument is working its way through the phases of unison shown on page 20.

If the variable pendulum is then further shortened in stages, a series of drawings like those opposite will be produced. The repetitive pattern represents beats with increasing frequency as the discrepancy between the notes widens. Eventually the series fades into a scribble that is a fair representation of discord, though even here there is a hint of some higher number pattern.

For most people this fading of visual harmony occurs at about the same point as the audible harmonies fade.

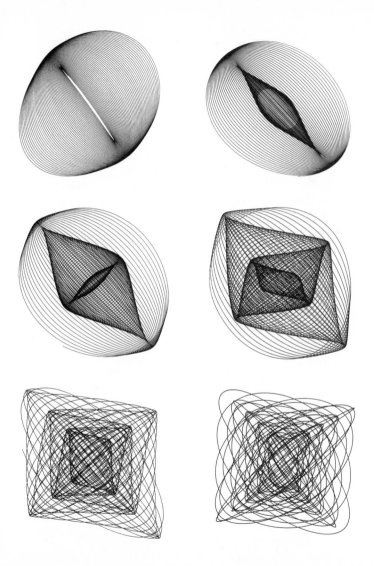

ROTARY UNISON—1:1
eggs and shells

Unison in contrary motion produces a straight line across the paper, like the closed phase of lateral unison. From concurrent motion there comes a mere dot that turns into a line struggling toward the center, pen and paper going around together.

At first this is disappointing. However, changing to near unison is richly rewarding. In contrary motion come a variety of beautiful, often shell-like, forms with fine cross hatchings. For best results lift the pen off the paper well before the pendulums reach equilibrium.

Surprisingly, from concurrent near-miss motion there come various spherical or egg-shaped forms. To produce an egg shape the pen should be lowered when it is dawdling at the center. It then spirals its way outwards, reaching a limit before returning as the pendulums run down. Because the lines toward the perimeter get closer together, the drawing appears three dimensional.

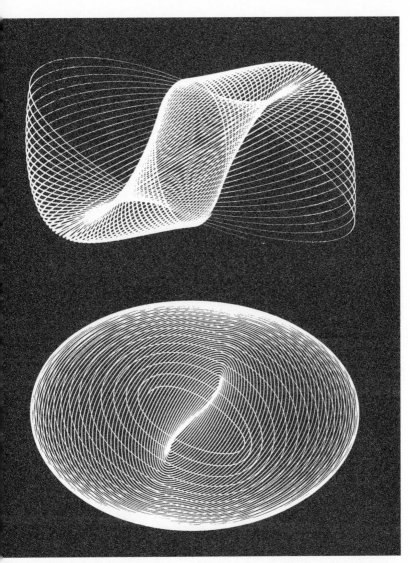

THE LATERAL OCTAVE—2:1
figure eights and wings

After unison the next harmony to try is the octave. Here there is a technical difficulty: The variable pendulum has to be very short, and because of the greater amount of friction involved it runs down quickly. The trick is to add a weight to the top of the invariable pendulum, which slows it down (*see title page*). The variable pendulum can then be longer.

Unfortunately this means that for the octave, and other ratios where one pendulum is going much faster than the other, the theoretical markers have to be ignored, and the right point found by trial and error.

With one pendulum beating twice as fast and at right angles to the other, the octave in open phase takes the form of a figure eight (a coincidence), repeated in diminishing size as the pendulum runs down.

If both pendulums are released at the same time to produce the closed phase, the result is a cup-shaped line that develops into a beautiful winged form with fine cross-hatchings and interference patterns. Small adjustments produce striking variations.

The octave is the first overtone (*see page 8*).

26

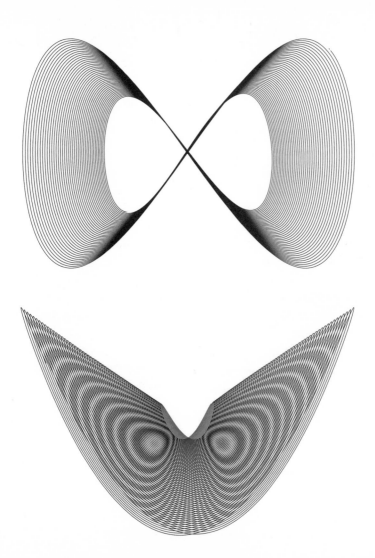

THE ROTARY OCTAVE—2:1
hearts and triangles

From rotary motion with a 2:1 ratio come some of the most beautiful of all harmonograph drawings: simple, graceful, and often surprising. Remember, all that is happening here is that two circular motions, one almost exactly twice as fast as the other, are being combined.

Contrary motion produces a trefoil shape with many fine variations (*opposite, right*). Starting with a smaller size or amplitude in the faster rotation produces a triangle, or pyramid.

The octave in concurrent motion produces a heart-shaped form with a simple inner loop (*below, left and opposite, left*). Here there is a link with the ancient tradition of the music of the spheres, for this is the shape an observer on Uranus would ascribe to the movement of Neptune, or vice-versa. This is because the planets orbit the Sun concurrently, Uranus in 84 years and Neptune in 165, approximately representing an octave.

Near misses in the ratios of rotary drawings set the designs spinning (*opposite, bottom*).

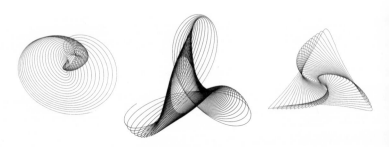

THE LATERAL FIFTH—3:2
and the second overtone 3:1

Next to be tried is the harmony of the fifth, intermediate between the simplicity of unison and octave and the more complex harmonies that follow.

It will be seen from the open phase drawing opposite that the fifth has three loops along the horizontal side and two along the vertical. The number of loops on each side gives the ratio, 3:2. Looking back at the octave, there are two loops to one, and with unison there is only one loop, however you look at it. This is the general rule for all lateral harmonograph ratios, and if a harmony appears unexpectedly during experiments, it can usually be identified by counting the loops on two adjacent sides.

The fifth also appears as 3:1, the second overtone, a fifth above the octave (*see open- and closed-phase drawings of 3:1 on page 3*). Drawing ratios outside the octave may require a twin-elliptic harmonograph (*see page 58*). The phase-shifted pair below are stereographic: if you go cross-eyed they will appear three dimensional.

31

THE ROTARY FIFTH—3:2
encircled hearts and fives

The loudness of musical tones is represented on the harmonograph by *amplitude*, the relative sizes of the two circular motions. In rotary drawings this is much more important than phase, which simply orients the whole design on the page.

The third drawing below shows a rotary fifth in contrary motion where the higher-frequency, faster-moving pendulum has a much wider swing. In the spiky drawing to its right it is the other way around. At equal amplitude all lines pass through the center *(see table on page 55)*.

The top four drawings opposite show rotary forms of 3:2, concurrent on the left, and countercurrent on the right. The second row shows the effect of a near miss in the harmony, which makes the patterns spin.

The lower two images opposite are of the second overtone, 3:1, a fifth above the octave (3:1 = 2:1 x 3:2). The concurrent version is on the left, countercurrent on the right.

With concurrent pictures, the number of swirls in the middle is given by the difference between the two numbers of the ratio. So the concurrent patterns for the primary musical intervals 2:1, 3:2, 4:3, 5:4, and 5:6 all have a single heart at their center.

THE FOURTH—4:3
with thirds, sixths, and sevenths

By now it will be evident that each harmony displays its own distinct aesthetic character. Unison is simple and assertive. The octave introduces an emphatic flourish, and the fifth, while still fairly simple, has added elegance.

With the fourth the pattern becomes more complicated, though the design is still recognizable without counting the loops. The upper diagram opposite shows the fourth in open phase, the lower in closed phase. An increasing sophistication becomes apparent, and some of the closed phase and near-miss variants have a strange, exotic quality.

Introducing the perfect thirds of diatonic tuning increases the complexity. The major third (5:4) is found below the fourth, the interval between them, a *diatonic halftone*, working out as $4:3 \div 5:4 = 16:15$. A fourth and a major third ($4:3 \times 5:4$) produce the major sixth, 5:3, a minor third (6:5) below the octave and a minor whole tone (10:9) above the fifth. Likewise, a fourth and a minor third ($4:3 \times 6:5$) create the minor sixth (8:5), a major third (5:4) below the octave and a halftone (16:15) above the fifth.

A fifth and a major third ($3:2 \times 5:4$) produce the major seventh, 15:8, while a fifth and a minor third ($3:2 \times 6:5$) give the minor seventh, 9:5. These are the elements of the diatonic, or *just*, scale.

FURTHER HARMONICS
seven-limit and higher-number ratios

As the numbers in the ratios increase it becomes harder to distinguish the harmonies one from another at a glance: The loops have to be counted, and slight variations produce little of aesthetic value. A typical example, 7:5, is shown opposite top.

Rotary motion produces a series of increasingly complex drawings, influenced by relative frequency, amplitude, and direction. In contrary motion the total number of loops equals the sum of the two numbers of the ratio. With concurrent motion the nodes turn inward, and their number is equal to the difference between the two numbers of the ratio.

The contrary drawings below show a fourth (4:3), another fourth, a major sixth (5:3) and a major third (5:4). The pictures opposite show unequal amplitude drawings of the perfect eleventh 8:3 (an octave and a fourth) and the ratio 7:3 which is found in seven–limit tuning (not covered in this book).

Two octaves and a major third (4:1 x 5:4) equal 5:1, the fourth overtone, which differs from four fifths $(3:2)^4$ as 80:81, the syntonic comma (*see page 10*). In *mean tone* tuning, popular during the Renaissance, the fifths were flattened very slightly, to $5^{1/4}$ or 1.4953, falling out of tune to please the thirds and sixths.

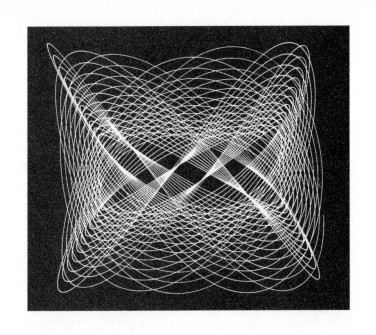

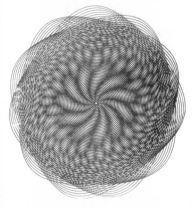
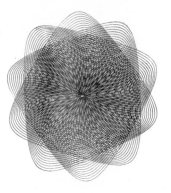

AMPLITUDE

circles, polygons, flowers, and another circle

Much variation can be obtained from a rotary ratio by having unequal sizes in the two circular motions. Opposite we see two frequencies related by a major sixth (5:3). A lower-frequency note begins to be influenced by, combines with, and is then more or less replaced by a higher-frequency one. When the two notes are at equal volume the lines all pass through the center (*see pages 56–57*). Notice that the sequence is not symmetrical.

Below we see the first three overtones. For the spikiest shapes simply invert the amplitudes. For polygons, square them first.

If you have ever played with a spirograph, the harmony is determined by the cogging ratio, and it is the amplitude that is adjusted when you change penholes on the wheel.

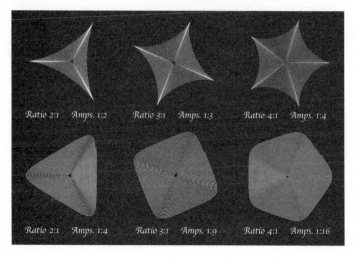

Ratio 2:1   Amps. 1:2   *Ratio 3:1   Amps. 1:3*   *Ratio 4:1   Amps. 1:4*

Ratio 2:1   Amps. 1:4   *Ratio 3:1   Amps. 1:9*   *Ratio 4:1   Amps. 1:16*

TUNING TROUBLES
the Pythagorean comma

Leaving the harmonograph drawings and returning to the principles of music, you may have noted that musical intervals do not always agree with one another. A famous example of this is the relationship between the octave and the perfect fifth (3:2).

In the central picture opposite, a note is sounded in the middle at 0, and moved up by perfect fifths (*numbered opposite, each turn of the spiral representing an octave*). After twelve fifths it has gone up seven octaves, but the picture shows that it has overshot the final octave slightly, and gone sharp. This is because $(3/2)^{12}$ ≈ 129.75, whereas $(2)^7 = 128$. The difference is known as the Pythagorean comma, 1.013643—approximately 74:73.

If you kept on spiraling you would eventually discover, as the Chinese did long ago, that 53 perfect fifths (or *Lü*) almost exactly equal 31 octaves. The first five fifths produce the pattern of the black notes on a piano, the *pentatonic* scale (*see page 50*).

The smaller pictures opposite show repeated progressions of the major third (5:4), the minor third (6:5), the fourth (4:3), and the whole tone (9:8) all compared to an invariant octave.

It's strange. With all this harmonious interplay of numbers you would have expected the whole system to be a precisely coherent whole. It isn't. There are echoes here from the scientific view of a world formed by broken symmetry, subject to quantum uncertainty, and (so far) defying a precise comprehensive theory of everything. Is this why the near miss is so often more beautiful than perfection?

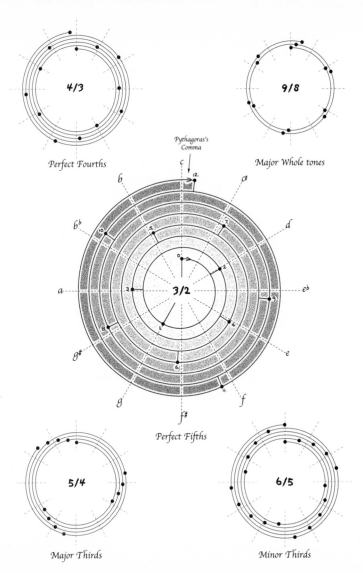

4/3

Perfect Fourths

9/8

Major Whole tones

Pythagoras's Comma

3/2

Perfect Fifths

5/4

Major Thirds

6/5

Minor Thirds

41

EQUAL TEMPERAMENT
changing keys made easy

Although early tunings enabled many pure harmonic ratios to be played, it was often hard to move into other keys; all one could do easily was change *mode* (*see page 52*). Musicians often had to retune their instruments, or use extra notes reserved for specific scales (classical Indian tuning uses twenty-two notes).

In the sixteenth century a new tuning was developed that revolutionized Western music and that predominates today. The octave is divided into twelve equal intervals, each *chromatic semitone* being 1.05946 times its neighbor ($2^{1/12}$, roughly 18:17).

Twelve equally spaced notes are arranged in a circle below. Six (flat) whole tones now make an octave, as do four (very flat) minor thirds, or three (sharp) major thirds. The Pythagorean comma vanishes, as do all perfect intervals except the octave. It's a clever fudge, slightly out of tune and we hear it every day.

Triads are chords of three notes. Opposite top we see major and minor triads involving the note C, in the key of C. Use the master grid (*opposite below*) to navigate the even-tempered sea, and place any triad in three distinct keys (*after Malcolm Stewart*).

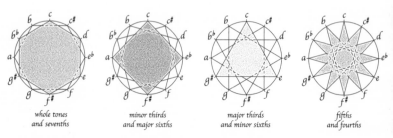

whole tones
and sevenths

minor thirds
and major sixths

major thirds
and minor sixths

fifths
and fourths

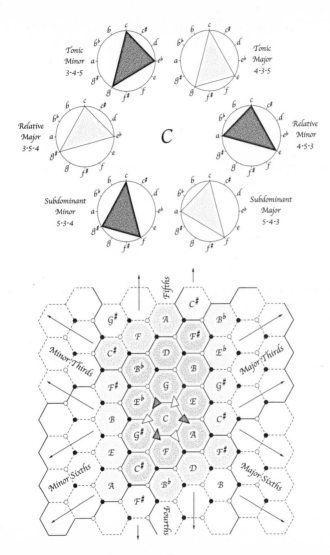

43

THE KALEIDOPHONE
squiggles from a vibrating rod

Despite the invention of equal temperament, scientists continued to investigate pure ratio harmonics. An interesting nineteenth-century precursor to the harmonograph was the kaleidophone, invented by Sir Charles Wheatstone in 1827. Like the harmonograph, it displayed images of harmonics.

The simplest version of the device consists of a steel rod with one end firmly fixed into a heavy brass stand and the other fixed to a small silver glass bead, so that when illuminated by a spotlight a bright spot of light is thrown up on a screen placed in front of it. Depending on how the kaleidophone is first struck, and then subsequently stroked with a violin bow, a surprising number of patterns can be produced (*a few are shown opposite*).

The kaleidophone does not behave like a string, as it is only fixed at one end. Like wind instruments, which are normally open at one end, the mathematics of its harmonics and overtones are more complicated than the monochord or the harmonograph (*the lower images opposite show some early overtones*).

Other versions of the kaleidophone used steel rods with square or oval cross sections to give further patterns. Wheatstone used to refer to his invention as a "philosophical toy," and indeed, as we look at these patterns, it is easy to feel wonder at their simple beauty.

To make your own kaleidophone, try fixing a knitting needle in a vise and attaching a silver bead to the free end. Use or make a light source that projects a bright point of light.

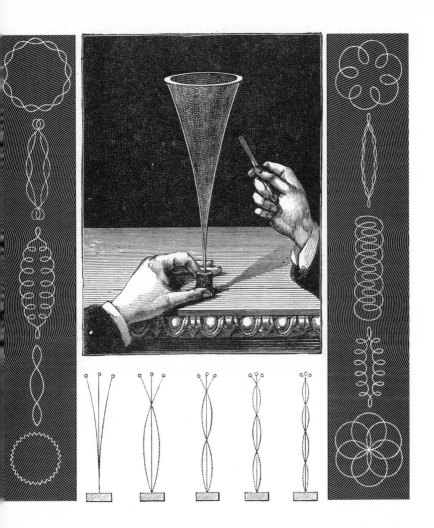

CHLADNI PATTERNS
vibrating surfaces

So far we have only considered vibrating strings and other simple systems, but surfaces also can be made to vibrate, and they too can display harmonic or resonant patterns.

In 1787 Ernst Chladni found that if he scattered fine sand onto a square plate, and bowed or otherwise vibrated it, then certain notes, generally harmonics of each other, each gave rise to different patterns in the sand on the plate. Like the harmonograph, other disharmonic tones produced a chaotic mess. Sometimes he found that further patterns could be created by touching the side of the plate at harmonic divisions of its length (*shown below*). This created a stationary node (*see page 8*). Later work revealed that circular plates gave circular patterns, triangular plates triangular patterns and so on.

The six pictures opposite are from Hans Jenny's book *Cymatics*, one of the seminal texts on this subject. The vibration picture appears gradually, the sand finding its way to the stationary parts of the plate as the volume increases.

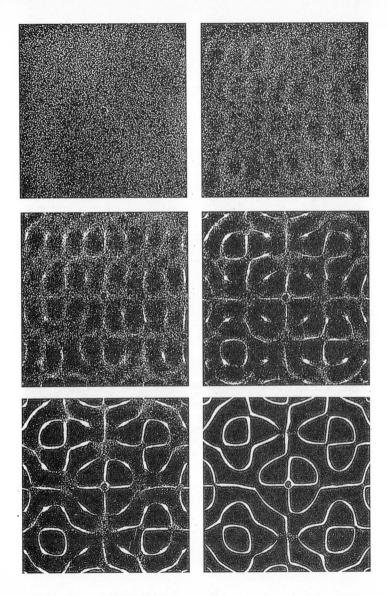

RESONANCE PICTURES
and how to sing a daisy

A more complete set of Chladni figures is shown opposite, all two- or fourfold because they were produced on a square plate.

Below, however, we see some circular patterns. They were photographed in the 1880s by Margaret Watts Hughes, a singer, on an ingenious device called an eidophone, which consisted of a hollow base with a membrane stretched across it and a tube attached to its base with a mouthpiece at the other end. As Mrs. Hughes sang diatonic scales down the tube, fine lycopodium powder scattered on the taut membrane suddenly came to life, bouncing away from some places and staying still at others, producing shapes that she likened to various flowers.

Yet again, we see recognizable forms and shapes appearing from simple resonance and harmony.

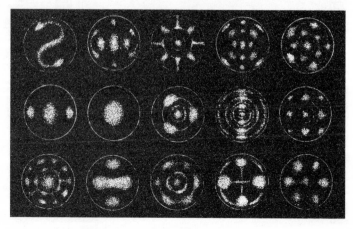

APPENDIX A: TUNINGS AND INTERVALS

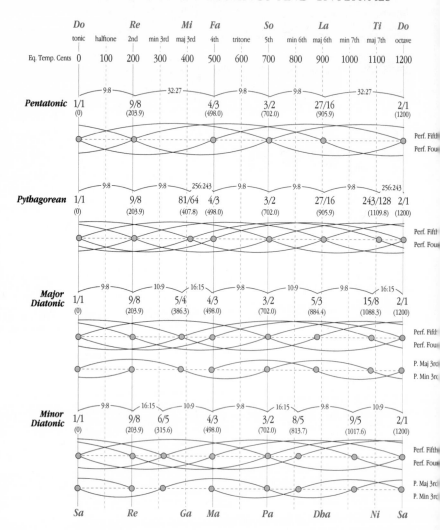

SELECTED MUSICAL INTERVALS

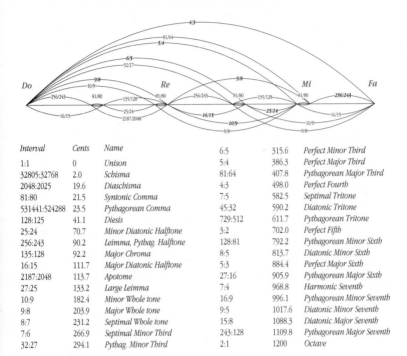

Interval	Cents	Name
1:1	0	*Unison*
32805:32768	2.0	*Schisma*
2048:2025	19.6	*Diaschisma*
81:80	21.5	*Syntonic Comma*
531441:524288	23.5	*Pythagorean Comma*
128:125	41.1	*Diesis*
25:24	70.7	*Minor Diatonic Halftone*
256:243	90.2	*Leimma, Pythag. Halftone*
135:128	92.2	*Major Chroma*
16:15	111.7	*Major Diatonic Halftone*
2187:2048	113.7	*Apotome*
27:25	133.2	*Large Leimma*
10:9	182.4	*Minor Whole tone*
9:8	203.9	*Major Whole tone*
8:7	231.2	*Septimal Whole tone*
7:6	266.9	*Septimal Minor Third*
32:27	294.1	*Pythag. Minor Third*

Interval	Cents	Name
6:5	315.6	*Perfect Minor Third*
5:4	386.3	*Perfect Major Third*
81:64	407.8	*Pythagorean Major Third*
4:3	498.0	*Perfect Fourth*
7:5	582.5	*Septimal Tritone*
45:32	590.2	*Diatonic Tritone*
729:512	611.7	*Pythagorean Tritone*
3:2	702.0	*Perfect Fifth*
128:81	792.2	*Pythagorean Minor Sixth*
8:5	813.7	*Diatonic Minor Sixth*
5:3	884.4	*Perfect Major Sixth*
27:16	905.9	*Pythagorean Major Sixth*
7:4	968.8	*Harmonic Seventh*
16:9	996.1	*Pythagorean Minor Seventh*
9:5	1017.6	*Diatonic Minor Seventh*
15:8	1088.3	*Diatonic Major Seventh*
243:128	1109.8	*Pythagorean Major Seventh*
2:1	1200	*Octave*

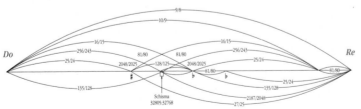

Like the off-center division of the octave into fifths and fourths, sharps are not in fact flats, giving rise to five more notes, making seventeen in all (found in Middle-Eastern tunings). More completely, we may think of the seven notes of the scale as moving across twelve "regions" of the octave, falling into the twenty-two positions of Indian tuning.

APPENDIX B: MODES AND EQUATIONS

Modern Names	The Seven Modes of Antiquity		Ancient Greek Names
Ionian *Major*	1 — 1 — ½ — 1 — 1 — 1 — ½ do re mi fa so la ti do	c d e f g a b c 1 2 3 4 5 6 7 8	**Lydian**
Dorian	1 — ½ — 1 — 1 — 1 — ½ — 1 re mi fa so la ti do re	d e f g a b c d 1 2 3♭ 4 5 6 7♭ 8	**Phrygian**
Phrygian	½ — 1 — 1 — 1 — ½ — 1 — 1 mi fa so la ti do re mi	e f g a b c d e 1 2♭ 3♭ 4 5 6♭ 7♭ 8	**Dorian**
Lydian	1 — 1 — 1 — ½ — 1 — 1 — ½ fa so la ti do re mi fa	f g a b c d e f 1 2 3 4♯ 5 6 7 8	**Syntolydian**
Myxolydian	1 — 1 — ½ — 1 — 1 — ½ — 1 so la ti do re mi fa so	g a b c d e f g 1 2 3 4 5 6 7♭ 8	**Ionian**
Aeolian *Natural Minor*	1 — ½ — 1 — 1 — ½ — 1 — 1 la ti do re mi fa so la	a b c d e f g a 1 2 3♭ 4 5 6♭ 7♭ 8	**Aeolian**
Locrian	½ — 1 — 1 — ½ — 1 — 1 — 1 ti do re mi fa so la ti	b c d e f g a b 1 2♭ 3♭ 4 5♭ 6♭ 7♭ 2	**Myxolydian**

The white notes on a piano give the seven notes of the seven modes of ancient Greece. Medieval transcription errors have left us with modern names that don't fit the ancient ones. Each mode, or scale, has its own pattern of whole tones and halftones, only two surviving as our major and natural minor scales.

Other scales include modal pentatonics that forbid semitones, the harmonic minor with its minor 3rd and 6th, 1 2 3♭ 4 5 6♭ 7 8, and many others.

The ratios and intervals in this book concern frequencies, normally expressed as cycles per second, or Hertz. Classical tuning sets C at 256 Hz. Modern tuning is higher, fixing A at 440 Hz. The period T of a wave is the reciprocal of its frequency f: $T = 1/f$.

The speed of sound in dry air is roughly $331.4 + 0.6\,T_c$ m/s, where T_c is the temperature in degrees celsius. Its value at room temperature, 20°C, is 343.4 m/s.

Gravitational acceleration on Earth, g, is 9.807 m/s^2.

Frequency of a pendulum	$\dfrac{1}{2\pi}\sqrt{\dfrac{gravitational\ acceleration}{pendulum\ length}}$
Fundamental frequency of a tensioned string	$\dfrac{1}{2 \times string\ length}\sqrt{\dfrac{string\ tension}{string\ mass \div string\ length}}$
Resonant frequency of a cavity with an opening	$\dfrac{speed\ of\ sound}{2\pi}\sqrt{\dfrac{area\ of\ opening}{volume\ of\ cavity \times length\ of\ opening}}$
Fundamental frequency of an open pipe or cylinder	$\dfrac{speed\ of\ sound}{2 \times length\ of\ cylinder}$

The beat frequency between f_1 and f_2 is the difference between them, $f_b = f_2 - f_1$.

The ratio a:b converts to cents (where a > b): $(\log(a)-\log(b)) \times (1200 \div \log 2)$. To convert cents into degrees multiply by 0.3.

Clapping in front of a rise of steps produces a series of echoes with a perceived frequency equal to $v/2d$, where v is the speed of sound, and d is the depth of each step. Clapping in a small corridor width w produces a frequency v/w.

The *arithmetic* and *harmonic means* are central to Pythagorean number theory. The arithmetic mean of two frequencies separated by an octave produces the fifth between them (3:2), the harmonic mean producing the fourth (4:3).

6	:	8	::	9	:	12
A	:	$\dfrac{A+B}{2}$	::	$\dfrac{2AB}{A+B}$	:	B
A	:	Arithmetic Mean	::	Harmonic Mean	:	B

Appendix C: Tables of Patterns

Overtone and simple ratio harmonics are shown below and opposite, arranged in order of increasing dissonance down the page. Open phase drawings display their ratio as the number of loops counted across and down. To find the ratio of a rotary drawing, draw both forms, concurrent (both circles in the same direction) and contrary (in opposite directions). Count the loops in each, add the two numbers together and divide the total by two. This gives the larger ratio number. Subtract this from the contrary total to give the lower ratio number. Rotary figures for the ratio *a:b* will have *b–a* loops when both circles are concurrent, and *a+b* loops when they are contrary.

The designs shown here were all made with equal amplitudes.

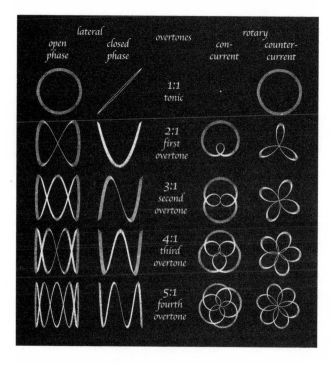

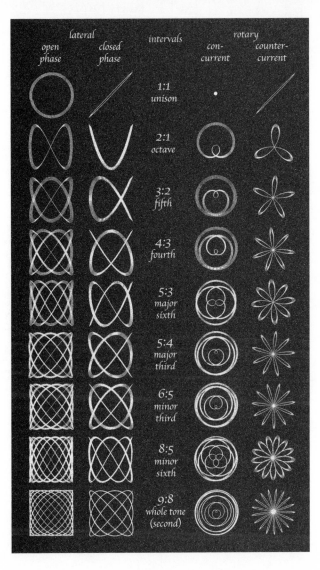

Appendix D: Building a
Harmonograph

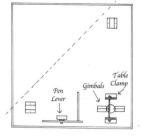

Pen Lever

Gimbals

Table Clamp

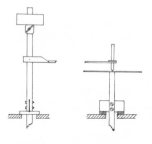

Anyone seriously interested in making a harmonograph should consider going straight for the three–pendulum model.

The table must be highly rigid and firm on the floor, otherwise the movements of the weights will be distorted. I suggest it should be about 36 inches (90 cm) above the floor, with a tabletop 24 x 12 inches (60 x 30 cm) for two pendulums, 24 x 24 inches (60 x 60 cm) for three, and about ¾ inch (2 cm) thick with an apron all round, about 3¼ inches (8 cm) deep.

The legs should be about 2½ inches (6 cm) square, splayed outward and pointed at the bottom. One way of achieving the splay is to fix wood or metal brackets in the corners under the table on each side of the diagonals and bolt the legs between them. After adjusting the legs to give the correct splay they can then be fixed in position with screws through the apron.

To save space, cut the tabletop as illustrated by the dotted line. Three legs are not quite as stable, but work fairly well.

The platform carrying the paper should be light and rigid, and fixed to the pendulum with a countersunk screw. Make the platform about 8⅔ x 6 inches (22 x 15 cm) to hold half an 8½ x 11-inch sheet secured by a rubber band or small clip.

All sizes suggested are maximum, but a scaled-down version will still work if it is carefully made.

If you are tempted to make a harmonograph, start with the weights, for the instrument will only be satisfactory if these are really heavy and yet easy to adjust. It is a good idea to make about ten, around 4 pounds (2 kilos) each, so the loadings can be varied. They should be about 3¼ inches (8 cm) in diameter, with a central hole, or with a slot for easier handling. Either cast them yourself from lead or ready-mixed cement or have them made by a metal shop or plumber.

The shafts should be made from wood dowel, about ½ inch (1.5 cm) in diameter (metal rods are liable to bend, distorting the drawings), marked off in inches.

Clamps can be obtained from suppliers of laboratory equipment. For some of the drawings top weights are needed, held in place by clamps. Clamps can also be added to pendulum tops for fine tuning, with one or more metal washers added.

The simpler kind of bearing consists of brass strips bolted into a slot in the pendulum and filed to sharp edges to rest in grooves on each side.

In a bearing involving less friction the pendulum is encased at the fulcrum in a horizontal block of hardwood with vertical bolts on each side filed to sharp points and resting in grooves in metal plates. If drilling the large hole in the block is too difficult, it can be made in two halves, each hollowed out to take the shaft and bolted together.

Rotary motion needs gimbals. Here the grooves for the pendulum are filed in the upper side of a ring (e.g. a key ring) while the under side has grooves at right angles to the upper ones. The lower grooves fit on two projecting sharp edges (brass strips), each enclosed between two pieces of wood fixed to the table. With the alternative bearing a large flat washer should be used with depressions to take the sharp points.

Pen arms should be as light as possible to minimize restraint. They are easily made from balsa wood strips (sold at model-making shops), using glue and Scotch tape. For two pendulums the arm can be fastened to the shaft with pinched-off needles, and the pen jammed into a hole at the other end. For three pendulums the side pieces on the arm should enclose its shaft firmly but not too tightly and be held gently with a thin rubber band. One of the arms holds the pen, while the other is held by protruding needles pushed in backward and secured (gently) at both ends by the rubber bands.

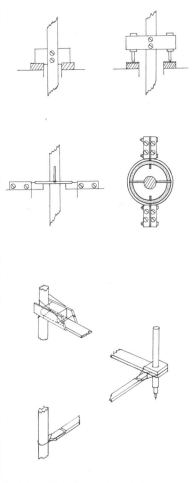

57

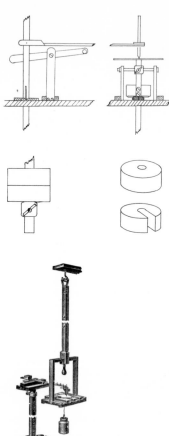

An additional fitting is needed to lock a rotary pendulum so that the instrument can be used with just the two single-axis pendulums. This can be done by mounting two brackets on the table near the rotary pendulum with holes to take a long horizontal bolt (slightly to one side) to which the shaft can be clamped.

Pens should be fine, light, and free-flowing. Most stationers and shops selling draftsmen's and artists' materials offer a variety (avoid ball point or thick fiber pens). For best results use shiny art paper and ordinary copier paper for preliminary experiments.

If the pen is left on the paper to the end there is usually an unsightly blob. To avoid this, mount a short pillar on the table with an adjustable lever carrying a piece of thin dowel placed under the pen arm. By raising the dowel gently the pen is lifted off the paper without jogging it. This device should also be used before the pen is lowered to the paper. By watching the pen you can see what pattern is being made, and nudge it one way or the other by pressure on the pendulums.

For ratios outside the octave, such as 4:1, you may need to try another harmonograph such as Goold's twin-elliptic pendulum (left).